DEVELOPING MATHEMATICS

**Customisable
teaching resources
for mathematics**

HANDLING DATA

Ages 8–9

Helen Glasspoole

A & C Black • London

Contents

Published 2009 by A & C Black Publishers Limited
36 Soho Square, London W1D 3HB
www.acblack.com

ISBN 978-1-4081-0040-0

Copyright text © Helen Glasspoole 2009
Copyright illustrations © Sean Longcroft 2009
Copyright cover illustration © Piers Baker 2009
Editors: Lynne Williamson and Marie Lister
Designed by Billin Design Solutions Ltd

The authors and publishers would like to thank
Catherine Yemm and Judith Wells for their advice
in producing this series of books.

A CIP catalogue record for this book is available from the
British Library.

Printed and bound in Great Britain

A & C Black uses paper produced with elemental chlorine-
free pulp, harvested from managed sustainable forests.

Introduction

100% New Developing Mathematics: Handling Data is a series of seven photocopiable activity books for children aged 4 to 11, designed to be used during the daily maths lesson. The books focus on the skills and concepts for Handling Data as outlined in the Primary National Strategy *Primary Framework for literacy and mathematics*. The activities are intended to be used in the time allocated to pupil activities in the daily maths lesson. They aim to reinforce the knowledge and develop the skills and understanding explored during the main part of the lesson, and to provide practice and consolidation of the learning objectives contained in the Framework document.

Handling Data

This strand of the *Primary Framework for mathematics* is concerned with helping pupils to develop the skills required to answer questions and solve problems by sorting and recording information according to certain given criteria and also criteria suggested by the children. In this strand they do this by recording in lists and tables and using practical resources, pictures, block graphs and pictograms. Broadly speaking, this strand addresses topic areas that were described under the 'Handling Data' strand title of the former National Numeracy Strategy *Framework for teaching mathematics*.

Handling Data Ages 8–9 supports the teaching of mathematics by providing a series of activities to develop understanding of key concepts within the handling data cycle. The activities provide opportunities for children to develop the skills of collecting, organising, presenting, analysing and interpreting data. The following objectives are covered:

- answer a question by identifying what data to collect; organise, present, analyse and then interpret the data in tables, diagrams, tally charts, pictograms and bar charts, using ICT where appropriate;
- compare the impact of representations where scales have intervals of differing step size.

Extension

Many of the activity sheets end with a challenge (**Now try this!**), which reinforces and extends children's learning, and provides the teacher with an opportunity for assessment. These might include harder questions, with numbers from a higher range, than those in the main part of the activity sheet. Some challenges are open-ended questions and provide opportunity for

children to think mathematically for themselves. Occasionally the challenge will require additional paper or that the children write on the reverse of the sheet itself. Many of the activities encourage children to generate their own questions or puzzles for a partner to solve.

Organisation

Very little equipment is needed, but it will be useful to have available: calculators, coloured pencils, counters, dice, scissors and glue. You will also need tape measures for page 14 and access to atlases, maps and the Internet for page 30. Blank Venn and Carroll diagrams, and blank vertical and horizontal bar charts are available on pages 59 to 62.

Where possible, children's work should be supported by ICT equipment, such as software for drawing tables and charts on interactive whiteboards, data logging equipment or spreadsheet packages for computers. It is also vital that children's experiences are introduced in real-life contexts and through practical activities. The teachers' notes at the foot of each page and the more detailed notes on pages 6 to 11 suggest ways in which this can be done effectively.

To help teachers select appropriate learning experiences for the children, the activities are grouped into sections within the book. However, the activities are not expected to be used in this order unless stated otherwise. The sheets are intended to support, rather than direct, the teacher's planning.

Some activities can be made easier or more challenging by masking or substituting numbers. You may wish to re-use pages by copying them onto card and laminating them.

Accompanying CD

The enclosed CD-ROM contains all of the activity sheets from the book and a program that allows you to edit them for printing or saving. This means that modifications can be made to further differentiate the activities to suit individual pupils' needs. See page 12 for further details.

Teachers' notes

Brief notes are provided at the foot of each page, giving ideas and suggestions for maximising the effectiveness of the activity sheets. These can be masked before copying.

Further explanations of the activities can be found on pages 6 to 11, together with examples of questions that you can ask. Answers can be found on pages 63 and 64.

Whole-class warm-up activities

The tools provided in A & C Black's *Maths Skills and Practice* CD-ROMs can be used as introductory activities for use with the whole class. In the *Maths Skills and Practice* CD-ROM 4 (ISBN 978-0-7136-7320-3), the Ice-cream Data activity could be used to introduce or reinforce 'Handling Data' objectives.

The following activities provide some practical ideas that can be used to introduce or reinforce the main teaching part of the lesson, or provide an interesting basis for discussion.

Fives

Give the children the opportunity to record and read data recorded in a tally chart. Ask the children to explain their strategies for calculating the number shown by the tally. Some might count on from the multiple of 5; others might think of a simple sum:

15 + 4

Sorted!

This activity reinforces the concept of sorting data using two criteria. Put labels on four clear plastic containers: two labels should be blank, one should say 'both' and the other 'neither'. The children then sort items into the containers. For example, if sorting a set of number cards into 'multiples of 3' and 'multiples of 2', the children write 'multiples of 3' on a sticky note and place it on one blank label and stick 'multiples of 2' on the other. They then take it in turns to pick a number card and put it in the correct container, explaining why it belongs there, for example: *This number belongs in the 'both' container because it is both a multiple of 2 and a multiple of 3; This number belongs in the 'neither' container because it is neither a multiple of 2 nor a multiple of 3.* The rest of the class can agree or disagree with the sorting. Look at the data and link to where they would be recorded on regions of Venn and Carroll diagrams.

Count me in

Practise counting on and back in intervals of 2, 5, 10 and 100. These skills will help children to interpret pictograms where one symbol represents more than one unit.

Practising scales

Mark a scale (using examples of both horizontal and vertical axes) in intervals of 2, 5 or 10. Can the children work out numbers half-way between the scales marked?

Missing numbers

Show examples of scales where some of the numbers are missing.

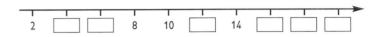

Working in pairs, the children discuss their ideas for the missing boxes. Invite them to think of a scale and complete part of it. Can others work out what the missing numbers are?

Notes on the activities

Answer a question by identifying what data to collect

The handling data cycle always begins with a question. It is important that children are given opportunities to consider how to answer a question. This involves deciding what information is needed and how to collect it. For example, when answering the question: 'What flavour of crisps should we have at the end-of-term party?', the children should work out for themselves that they need to find the class' favourite crisp flavours, and then decide how they will go about finding this out. This should include the children generating their own lists of flavours to use in a survey of the class. This will help the children begin to realise that the way a question is worded will determine what sort of information is needed to answer it, or even whether it can be answered at all.

Data detectives (page 13)

To reinforce the idea that different questions will result in different collection methods, this activity requires the children to think about the nature of different questions. This should build on examples that the children are familiar with, such as counting data, measuring data (for example temperature or speed), keeping a tally if the data cannot be counted, or collection of data through simple surveys. The focus is on children thinking about how to answer different lines of enquiry and not on actually collecting the data.

SUGGESTED QUESTIONS:

- How can you find the answer to the question?
- Is there more than one way to find the answer?

Dem bones (page 14)

The main focus of this activity is for the children to decide what data they need to collect. They also need to decide how they will collect it and how they will present their results. Encourage them to choose how they will record their findings before they start measuring. At the start of the lesson, you could ask the children to predict which bone they think is the longest, and why they think so. Children may suggest that the backbone is the longest bone and may be interested in the fact that the spine is made up of 33 vertebrae. Children may be interested to learn that there are about 206 bones in the adult skeleton, but that babies have about 300 bones – some of the bones fuse together as they grow up – so the children will have more bones than the teachers!

SUGGESTED QUESTIONS:

- What data do you need to collect?
- How will you collect the data?
- Why did you decide to record your result like that?

Organise, present, analyse and interpret data in tables and tally charts

Children will be used to organising data in lists and tables. Remind them of the process of tallying: making a vertical mark for each one, and then at the fifth count, crossing the previous four strokes through. Discuss the value of tallying as a way of recording a count and ask for examples of situations where tallying is useful, and occasions when simply counting results is easier.

Jelly belly: 1 and 2 (pages 15 and 16)

Discuss why the construction of tally charts is not appropriate here (the data can be counted easily). Before the children begin the activity, ask them to look at the table to get a sense of the different types of jelly.

SUGGESTED QUESTIONS:

- How could you check that you have recorded the data correctly?
- Can you use the table to work out which is the most popular flavour of jelly? Why not?

Rain recorder (page 17)

At the start of the lesson, explain how rainfall is measured using a rain gauge and point out that the data in the table has been rounded to the nearest centimetre. Ask the children to think about what the data does and does not show, using the suggested questions below. Some children might suggest that there could have been concentrated rainfall during part of the month and the rest of the month could have been hot and sunny. Although the focus is on interpretation of the table, the children could represent the data on a vertical or horizontal bar chart.

SUGGESTED QUESTIONS:

- How much rain fell in September?
- Does the data tell us whether it rained every day?
- What might have happened in October?

No shoes allowed! (page 18)

The data shows a tally of the number of pairs of different types of footwear taken off by children who had a go on a bouncy castle. Question 5b requires the children to contextualise the data and to draw conclusions about the weather on the day of the fête, supported by reasoning that flip-flops were a popular choice and that it was probably too hot to wear boots.

In the extension activity, the total number of pairs will give the total number of children who went on the bouncy castle, with the exception of the two children who turned up with bare feet. The children should, therefore, include the bare-footed children in the data to give an accurate result.

SUGGESTED QUESTIONS:

- How many children wore flip-flops/school shoes/trainers?
- What does a tally of ⅢⅠ ⅠⅠ show?

Tally ho! (page 19)

This activity focuses on the interpretation of a tally chart in order to answer questions. Ask the children to think about the context of the data and discuss which data they think should be included. Is it possible, for example, to tell from the tally chart which horse won the award for the best jumper?

SUGGESTED QUESTIONS:

- What questions would you like to ask about the horses?
- What does the data tell us about Indigo Blue?

Rock, paper, scissors (page 20)

Game rules The children should play the game in pairs. Both partners make a fist of one hand. Simultaneously, they gently punch the air in front of them three times and on the third punch they show a rock (keep as fist), paper (flat hand) or scissors (first two fingers open like a pair of scissors). The following shows who the winner of each pair is, including a rationale:

- rock beats scissors – it blunts them;
- scissors beat paper – they cut it;
- paper beats rock – it covers/wraps around it.

Emphasise that it is useful to keep a tally chart here because the data is collected over a period of time and needs to be counted at the end. This can be linked to the importance of keeping a correct score in any game situation.

SUGGESTED QUESTIONS:

- What sort of data is recorded using a tally chart?
- When might it be important to keep a tally chart?

Organise, present, analyse and interpret data in Venn diagrams and Carroll diagrams

The children will be aware that Venn and Carroll diagrams are used for sorting data. When working with either type of diagram, remind the children that all the data must be shown somewhere on the diagram. In a Venn diagram, data that does not fit either criterion is placed outside the rings but within the rectangle. In a Carroll diagram, there is a box for data that does not match either criterion.

The children might be interested to learn that Carroll diagrams are named after Lewis Carroll who wrote *Alice in Wonderland*, as he was the first to use them.

Hen diagram (page 21)

This activity reinforces children's understanding of sorting discrete data according to two criteria. They should understand the nature of the data in the intersection: the data belongs equally to both sets. Each area of the Venn diagram has been labelled to reinforce that different data belongs in different places.

SUGGESTED QUESTIONS:

- What data belongs in the left-hand/right-hand circle?
- What data belongs in the middle section?
- Can any new hen be included in the diagram?

Sorting symmetry (page 22)

Recap horizontal and vertical lines of reflective symmetry before the children begin this activity. Ensure they understand that the alphabet on the worksheet is made up of a mixture of capital and lowercase letters. The extension activity asks the children to complete the labels 'Symmetrical and capital' and 'Not symmetrical, not capital'. Some children might find the sorting easier if these labels are already filled in.

SUGGESTED QUESTIONS:

- What does 'symmetrical' mean?
- How can a mirror help you to decide whether a letter is symmetrical or not?
- How many of the letters are symmetrical and capital?

Monsters' tea party (page 23)

Ensure that the children understand how the two criteria are represented on the Carroll diagram. Look at diagrams showing one criterion to consolidate the concept, if necessary. The focus here is on the interpretation of the data in the diagram.

SUGGESTED QUESTIONS:

- How could we record someone who had lemonade and fruitcake?
- What did Wolfy have? How do you know?

2s and 3s (page 24)

Before starting the activity, revise the 2 times- and 3 times-tables. Ensure that the children understand the layout of the Carroll diagram, reinforcing the two criteria (multiple of 2, multiple of 3) and the sections which include both multiples of 2 and 3, and neither multiples of 2 nor 3.

SUGGESTED QUESTIONS:

- Is this number a multiple of 2? Is it a multiple of 3? Where should it be placed in the diagram?
- Which numbers should be placed in this section of the diagram?

Sweet success (page 25)

Before starting the activity, ensure that the children have had an opportunity to sort objects and to discuss the criteria by which they have sorted them. This activity requires the children to decide how they want to sort the sweets and to record the data on the relevant diagrams. Further ideas for sorting can be recorded on a separate piece of paper. As an extension, the children could explain to a partner how they sorted the sweets.

SUGGESTED QUESTIONS:

- How many different ways can you sort the sweets?
- Is there a way of sorting the sweets so that they are all included in one region of a Venn diagram?

Organise, present, analyse and interpret data in pictograms

Pictograms are a good introduction to recording results in graphs and charts, as the content is often very obvious through the use of symbols. Children will be used to interpreting and analysing pictograms where a symbol stands for one unit. It is important that they also recognise how symbols can be used to represent more than one unit, and how this changes the look of a chart.

Minibeasts (page 26)

This activity revises pictograms where one symbol represents one unit of data. Children could discuss why there does not need to be a numbered scale along the bottom of the diagram.

SUGGESTED QUESTIONS:

- What does the symbol of the magnifying glass mean?
- How many worms were found?

Sock sort (page 27)

The focus of this activity is careful interpretation of what the symbol represents on the pictogram. The symbol stands for two socks or one pair of socks. The questions assess whether the children have understood this. A link to the real-life context of losing a sock will give the children an opportunity to discuss why the data shows an odd number of socks.

SUGGESTED QUESTIONS:

- Why is the same symbol used for all the different types of sock?
- How many pairs of pink socks are there?
- How many pink socks are there altogether?

Summer fête (page 28)

This activity looks at a pictogram where each symbol represents ten units of data, with half a symbol representing five units of data. To consolidate understanding of the value of the symbol, the extension activity requires the children to use the symbol to represent a given number.

SUGGESTED QUESTIONS:

- What does the symbol represent?
- How many people would nine symbols represent?

Super Song Star (page 29)

This activity focuses on the value of each symbol. Each microphone represents 100 votes. The data reflects votes rounded to the nearest 50. Revise rounding if necessary.

SUGGESTED QUESTIONS:

- How many votes did The Magnets get?
- How many votes would six symbols represent?

Stamp about: 1 and 2 (pages 30 and 31)

If preferred, the sorting of the 'envelopes' could be carried out as whole-class research. Draw attention to the fact that the envelope symbol represents two letters.

SUGGESTED QUESTIONS:

- How many letters are going to places outside of the UK?
- How many letters are going to the rest of the world?

Mr Folly's lollies (page 32)

This activity requires the children to interpret a tally chart and then present a pictogram of the same data. The children choose how many units are represented by one symbol of the pictogram.

SUGGESTED QUESTIONS:

- How many lollies were sold on Saturday?
- How did you work this out?
- How many lollies are represented by one symbol?

Organise, present, analyse and interpret data in bar charts

Children will be used to interpreting and analysing bar charts where the scales go up in steps of one. It is important that they also recognise that scales can go up in intervals of more than one unit, and how this changes the look of a bar chart.

This section includes vertical and horizontal bar charts. Encourage the children to use both kinds and to remember that every bar chart must have a title and the axes must be labelled.

After-school sports (page 33)

At the start of the lesson, ask the children to say what they know about bar charts. Ask them to list everything that a bar chart should have on it.

SUGGESTED QUESTIONS:

- How many people took part in the survey?
- Looking at the bars on the bar chart, what can you say about the data?

Splish, splash (page 34)

This activity gives the children an opportunity to draw a bar chart where the scale is labelled in ones.

SUGGESTED QUESTIONS:

- How many babies were in the small pool on Wednesday?
- On which day were there 19 babies in the small pool?

Family zoo (page 35)

The main emphasis is to interpret data in a bar chart. The questions relating to the number of children who own each type of pet (calculated from data) and the number of children in the class (information given in a question) prompts the children to think about what the data is and how it was collected, i.e. some children might own many pets whilst others do not own any. Unlike data relating to, for example, eye colour or favourite colour, the total number of units of data does not indicate how many people took part in the survey. It is likely that some children in your class might own pets that are not shown on the chart – they could discuss this and think about whether all pets should be represented on a bar chart.

SUGGESTED QUESTIONS:

- What does the title of the bar chart tell you?
- Do you own any pets? Could you add that data to the bar chart?

What a smoothie! (page 36)

This activity involves the children drawing a bar chart and using a scale where the intervals are in twos. At the start of the lesson, practise counting in twos. Then draw a scale on the board where the numbering goes up in steps of two. Ask the children to say where, for example, 13, 23, 17, 5 would be shown on the scale.

SUGGESTED QUESTIONS:

- How did you know to end that bar there?
- Where is 9 shown on this scale?

Robot sort (page 37)

Here, the children fill in the empty labels on the bar chart and then draw in the two missing bars. They will need to look at the values in the table in conjunction with the number of divisions along the horizontal axis and the bars already completed when thinking about the scale.

SUGGESTED QUESTIONS:

- How many robots can spin?
- What is the highest value that is shown by the data?

Golden Giants (page 38)

The children use the data in a table to draw a vertical bar chart. Ensure that the children mark the sale in steps of two correctly, and check that they understand that a bar of height 13 will be halfway between the marks for 12 and 14 on the vertical axis. The data representations can then be evaluated.

SUGGESTED QUESTIONS:

- How many games did the Stingrays win?
- The Wild Cats won eleven games. Where would the end of their bar be on the bar chart?

Cool shades (page 39)

The data shows 30 pictures for children to sort according to shape of shades. The data can be counted (and crossed off to support data collection), and then recorded in the total boxes in preparation for presentation as a bar chart.

SUGGESTED QUESTIONS:

- How do you know how many of each shape of shades there are?
- Does it matter in which order you plot the bars on the bar chart?

Beachcomber (page 40)

This activity presents data in a table and requires the children to present it in a bar chart. The focus is on choosing an appropriate scale for the bar chart. Six equally-spaced intervals on the vertical axis guide the children in their decision making.

SUGGESTED QUESTIONS:

- What is the maximum value that will be shown on the bar chart?
- Would intervals of ten be a good choice for your scale? Why/why not?

Bounce 4 charity (page 41)

This activity focuses on the interpretation of a horizontal bar chart where the scale shows intervals of five. It also gives an opportunity for the children to consider what the data means, or could mean, by asking why they think that more money was raised by one particular class (suggestions are included in the answers).

SUGGESTED QUESTIONS:

- Which class raised £85?
- Which class raised £65?

Recycling (page 42)

Ensure the children understand that the data shows the weight of different materials collected over a month and that the focus is on comparison of different materials. The extension activity prompts the children to think about the context and meaning of the data; that because plastic is lighter than glass, it may appear that fewer plastic objects were collected, which may not have been the case.

SUGGESTED QUESTIONS:

- How many kilograms of glass were collected?
- Would a bar chart for the recycling done on your street be similar or different to this one?

Adventure holiday (page 43)

The main focus of this activity is for children to construct a horizontal bar chart using an appropriate scale.

SUGGESTED QUESTIONS:

- What scale do you think is the best one to use?
- Why do you think this?

Analyse and interpret data in tables, pictograms and bar charts

> It is important children realise that the same set of data can be shown in many different ways. This section provides opportunities for the children to look at data represented in different formats and to analyse the different graphs and charts shown.

Odd one out: 1 and 2 (pages 44 and 45)

Ensure that the children are familiar with a range of representations, including tally charts, tables, pictograms and vertical and horizontal bar charts. This activity requires the children to compare three different representations of the same data in order to identify which one is incorrect and, therefore, the odd one out. You could use this as an assessment activity or as a way of checking whether the children find reading particular diagrams difficult.

SUGGESTED QUESTIONS:

- What information will help you to spot the odd one out?
- In set B, how did you know that the pictogram was the odd one out?

Game, set and match: 1 and 2 (pages 46 and 47)

This activity is best done towards the end of the year to consolidate the children's handling data skills. The focus is on discussion, reasoning and interpretation of charts in order to match them to a simple scenario.

SUGGESTED QUESTIONS:

- If we drew a bar chart for B/D, how many bars would there have to be?
- Tell me another situation that could fit the data in D.

Missing data (page 48)

In the context of some data having been 'deleted', the children are required to piece together information from the table and from the bar chart to find the missing data. The amount of glass pieces (65) can be determined by subtracting the known amounts from the total. Children working in pairs will maximise opportunities for reasoning, communicating and problem solving.

SUGGESTED QUESTIONS:

- How many pieces of woven fabric were found?
- What was the total number of things that were found?

Top of the Mountain: 1 and 2 (pages 49 and 50)

This track game for two players (or more if preferred) is designed to consolidate concepts and skills linked to handling data. The answers can be discussed between players for an agreement to be reached, or a third child could act as adjudicator.

SUGGESTED QUESTIONS:

- Do you think that the answer is correct?
- How could you check?

Compare the impact of representations where scales have intervals of differing step size

During this year, children will become accustomed to recording data in and interpreting information from charts where the scales have intervals other than in ones. It is valuable for children to have opportunities to see how different scales on axes affect how the same data is presented, and to make a habit of always checking the scale before commenting on the information shown by a pictogram or bar graph.

Mix and match: 1 and 2 (pages 51 and 52)

The children should discuss their choice of pairs with a partner (or in a small group), giving explanations, and reasoning orally about their choices.

SUGGESTED QUESTIONS:

- What clues can you find to help you to work out which pairs match?
- How do you know that these two bar charts show the same data?

Scaly fish: 1 and 2 (pages 53 and 54)

In this activity, the children plot the same data on four different vertical bar charts, each of which has a different scale. They should first plot the data, then evaluate the representations and discuss which one they would choose and why. (This will lay the foundation for children to understand, at a later date, how data representations can be manipulated to tell a particular story in, for example, the media.)

SUGGESTED QUESTIONS:

- What is the total number of fish counted?
- Do you think that this is the total number of fish in Mika's aquarium?
- What makes you think this?

Handling data investigation

This section consists of eight worksheets to support a handling data investigation. The worksheets are presented at A5 size in the book and on the CD-ROM, but could be enlarged. Each step of the handling data cycle is developed as children follow a line of enquiry. Different stages of the cycle could be made a particular focus for each enquiry.

Data investigation (page 55a)

Use this worksheet to remind the children of the stages in the handling data cycle, and to record the members of the group and the question that the group would like to investigate. This question should be filled in once the group have completed page 55b (What's the problem?) and have come to a decision about what they would like to investigate. This worksheet can form the cover sheet for the investigation.

What's the problem? (page 55b)

As far as possible, encourage the children to think of a question that is meaningful to them, and to which they would like to find the answer. Examples could be linked to choice of school dinners, break-time and lunch-time activities, games to be painted on the playground, where the litter bins should be located, what fruit is popular for break time.

The opportunity for children to predict an outcome should not encourage them to look for this as the answer to the question that they have posed. Instead, it should help them to think about the nature of the data that they have to collect.

How will you collect your data? (page 56a)

The children decide which is the best method for collecting their data. It is important that they are able to explain why they think that their chosen method is appropriate. Make links to scientific ideas of a 'fair test' and support children in thinking about how many people they need to ask or places they should look and the practical implications of gathering the information they need.

Record your data (page 56b)

ICT could be used to create a table. The children can discuss the practical method of recording data. It might be appropriate for the group to gather data from different people and then pool the data at the next stage.

Organise your data (page 57a)

Before thinking about how they will represent their data, the children will need to organise the data. This might include counting the tally marks and completing a table, pooling data (where individuals collected part of the data), counting responses from surveys, etc.

Represent your data (page 57b)

Carry out a class discussion about the strengths of different representations. Then ask the groups to decide how they would like to represent their data. At this stage, and if appropriate, individuals within the group could represent the same data in different ways for later evaluation. It is important that the children can explain why they have chosen a particular representation, rather than spending time colouring in bar charts, etc. If more space is needed, the children could use an additional piece of paper. ICT could be used, and relevant learning objectives incorporated at this stage.

Interpret your data (page 58a)

This is an opportunity for the children to revisit their prediction and to talk about whether there were any surprising results. Writing questions about their data for other children to answer will prompt them to look at what their data means and to assess whether their representation is clear enough for others to understand.

Evaluate your investigation (page 58b)

In addition to thinking about whether the data answered their question, this worksheet gives children the opportunity to evaluate each step of their enquiry and to suggest ways to make future investigations easier or clearer. Link back to the first stage of the investigation by considering whether there are any follow-up questions or new questions to which the children wish to find the answers.

Resources

This section contains blank diagrams and graphs for children to use. They can be used in several ways: on the interactive whiteboard for whole-class activities, enlarged to A3 and used for sorting objects, copied for individuals or pairs to complete. They can be used: with data prepared by the teacher or by individuals or small groups, blank for children to collect and present data, to support activities in this book, as part of extension activities where children present the data from a worksheet in a different format.

Blank Venn diagram (page 59)

Blank Carroll diagram (page 60)

Blank vertical bar chart (page 61)

Blank horizontal bar chart (page 62)

Using the CD-ROM

The CD-ROM included with this book contains an easy-to-use software program that allows you to print out pages from the book, to view them (e.g. on an interactive whiteboard) or to customise the activities to suit the needs of your pupils.

Getting started

It's easy to run the software. Simply insert the CD-ROM into your CD drive and the disk should autorun and launch the interface in your web browser.

If the disk does not autorun, open 'My Computer' and select the CD drive, then open the file 'start.html'.

Please note: this CD-ROM is designed for use on a PC. It will also run on most Apple Macintosh computers in Safari however, due to the differences between Mac and PC fonts, you may experience some unavoidable variations in the typography and page layouts of the activity sheets.

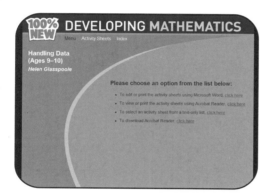

The Menu screen

Four options are available to you from the main menu screen.

The first option takes you to the Activity Sheets screen, where you can choose an activity sheet to edit or print out using Microsoft Word.

(If you do not have the Microsoft Office suite, you might like to consider using OpenOffice instead. This is a multi-platform and multi-lingual office suite, and an 'open-source' project. It is compatible with all other major office suites, and the product is free to download, use and distribute. The homepage for OpenOffice on the Internet is: www.openoffice.org.)

The second option on the main menu screen opens a PDF file of the entire book using Adobe Reader (see below). This format is ideal for printing out copies of the activity sheets or for displaying them, for example on an interactive whiteboard.

The third option allows you to choose a page to edit from a text-only list of the activity sheets, as an alternative to the graphical interface on the Activity Sheets screen.

Adobe Reader is free to download and to use. If it is not already installed on your computer, the fourth link takes you to the download page on the Adobe website.

You can also navigate directly to any of the three screens at any time by using the tabs at the top.

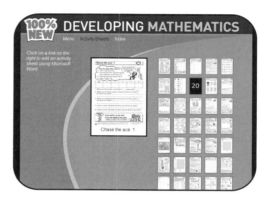

The Activity Sheets screen

This screen shows thumbnails of all the activity sheets in the book. Rolling the mouse over a thumbnail highlights the page number and also brings up a preview image of the page.

Click on the thumbnail to open a version of the page in Microsoft Word (or an equivalent software program, see above.) The full range of editing tools are available to you here to customise the page to suit the needs of your particular pupils. You can print out copies of the page or save a copy of your edited version onto your computer.

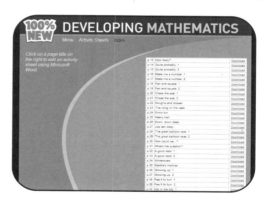

The Index screen

This is a text-only version of the Activity Sheets screen described above. Choose an activity sheet and click on the 'download' link to open a version of the page in Microsoft Word to edit or print out.

Technical support

If you have any questions regarding the *100% New Developing Literacy* or *Developing Mathematics* software, please email us at the address below. We will get back to you as quickly as possible.

educationalsales@acblack.com

Data detectives

- **Look at the questions asked by these children.**

a Who in the class is the fastest at running?

b How many people will walk their dogs past school today?

c Which is the hottest part of the room you are in?

d Are there more children who are aged 8 or aged 9 in the class?

e Which is the most popular computer game in the class?

- **What data would you suggest they collect to answer their questions and how? Record your ideas below.**

a _____ ruoms _____

b _____ gair _____

c _____ my kitchen _____

d _____ aged 9 _____

e _____ mario _____

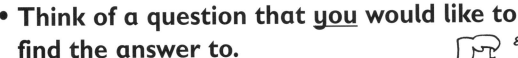

NOW TRY THIS!

- **Think of a question that <u>you</u> would like to find the answer to.**
- **How could you find out the answer?**
- **Talk to a partner about your ideas.**

Teachers' note This activity offers a good opportunity for discussion between children. Ask them to say why they would collect data in a certain way and encourage them to evaluate their ideas alongside the suggestions of others.

100% New Developing Mathematics
Handling Data: Ages 8–9
© A & C BLACK

Dem bones

- ## What is the longest bone in your body?

How could you find out?

Work with a partner.

What data do you need to collect?

How will you collect it?

What equipment do you need?

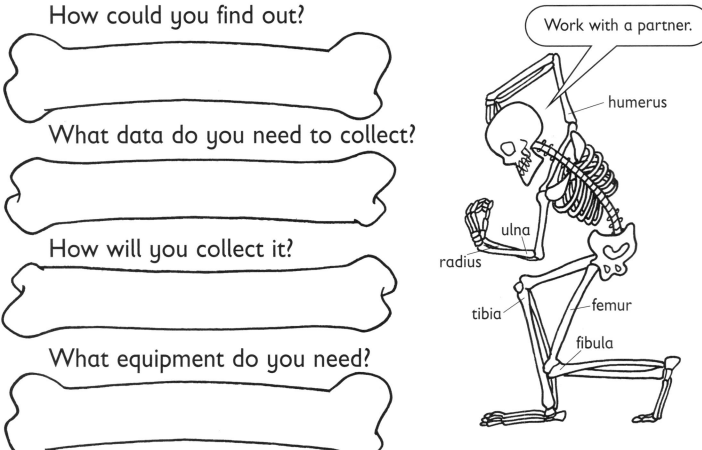

humerus

ulna

radius

tibia

femur

fibula

- ## Choose how you want to present your results:

NOW TRY THIS!

- ## Is your answer the same as everyone else's in your class? How could you find out?

Teachers' note Tape measures (one per pair) will be needed for this activity. Read through the names of the bones as a class to ensure that the children know what they are measuring. The children can choose how they want to record their results in the box or on a separate sheet of paper. Whole-class data could be collected and represented on the board.

100% New Developing Mathematics
Handling Data: Ages 8–9
© A & C BLACK

Jelly belly: 1

Lots of different shapes of jelly were made for the Jelly Eating Competition.

You need a copy of Jelly belly: 2.

- **Look at the data on Jelly belly: 2.**
- **Count how many jellies of each shape there were at the start of the competition.**
- **Record the information in the table below.**

Type of jelly		Number of jellies
Rabbit jelly		4
Wobbly jelly		
Brain jelly		
Pumpkin jelly		
Bear jelly		
Total number of jellies		

NOW TRY THIS!

There are three different flavours of jelly.

- **Count how many jellies of each flavour there are and record here:**

◯ Lime	◔ Strawberry	● Blackcurrant

Teachers' note Use in conjunction with page 16, Jelly belly: 2. Each category of jelly should be counted and the totals recorded in the table. The jellies are shaded to show three different flavours for the extension activity. The children could make this clearer using coloured pencils.

100% New Developing Mathematics
Handling data: Ages 8–9
© A & C BLACK

Jelly belly: 2

Teachers' note Use in conjunction with page 15, Jelly belly: 1.

100% New Developing Mathematics
Handling Data: Ages 8–9
© A & C BLACK

Rain recorder

The Isle of Wight is an English island off the south coast of England.

The table below shows typical rainfall on the island, rounded to the nearest centimetre.

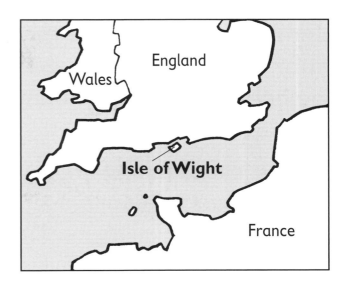

January	February	March	April	May	June
8 cm	5 cm	7 cm	5 cm	9 cm	5 cm

July	August	September	October	November	December
6 cm	8 cm	5 cm	13 cm	7 cm	9 cm

• **Use the table to answer these questions.**

1 How much rain fell in August? ____ cm

2 In which months was 7 cm of rain collected?

_____ and _____

3 How many months had a rainfall of 5 cm? ____

4 a In which month did the most rain fall? _____

b How much more rain fell in this month than in May? ____ cm

5 How much rain fell in the month of your birthday? ____ cm

NOW TRY THIS!

• **How much rain was collected over the whole of the year?** ____ cm

Teachers' note This data is taken from http://www.isleofwightweather.com/ The data is rounded to the nearest centimetre. Calculators could be made available for the extension activity.

100% New Developing Mathematics
Handling data: Ages 8–9
© A & C BLACK

No shoes allowed!

Jemima was asked to make sure that children took off their footwear before going on the bouncy castle.

1 Complete the 'total' column in the **tally chart** to show how many pairs of each type of footwear were recorded during the time Jemima was in charge.

Type of footwear	Tally (number of pairs)	Total
Flip-flops	𝍸 𝍸 𝍸 ||	17
Plimsolls	𝍸 ||	
Trainers	𝍸 𝍸 𝍸 𝍸 |||	
Boots	|	
Wellies	||||	
School shoes	||	

2 How many children wore wellies? ____

3 How many pairs of flip-flops were there? ____

4 There were **exactly** seven pairs of which type of footwear?

____

5 a Which type of footwear did **fewest** children wear? ____________

b Why might this be? ______________________________

NOW TRY THIS!

Two children came in bare feet.

- **Add these bare-footed children to the tally chart.**
- **How many children in total went on the bouncy castle whilst Jemima was in charge?** ____

Teachers' note If necessary, revise how to record using a tally system. Explain to the children that each child has one go only on the bouncy castle.

Tally ho!

The local horse club held a competition.
This tally chart shows how many times each horse
jumped successfully over a different fence.

Name of horse	Number of successful jumps	Total
Chestnut	ЖЖ II	7
Dapple	ЖЖ ЖЖ II	
Indigo Blue	IIII	
Jasper	ЖЖ ЖЖ ЖЖ	
White Knight		
Oyster Pearl	ЖЖ ЖЖ I	

1 In the chart, record the total number of jumps for each horse.

2 a Which horse jumped successfully over the most fences?

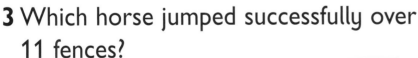

b How many more fences than Indigo Blue
did this horse jump?

3 Which horse jumped successfully over
11 fences? _____

4 What do you think happened to White Knight?

_____

NOW TRY THIS!

**Fen Star had three more successful
jumps than Indigo Blue.**
• Show Fen Star's total as a tally. _____

Teachers' note The children could work in pairs or small groups to support discussion, particularly
for question 4. Encourage the children to think about what the data does and does not tell them.

100% New Developing Mathematics
Handling data: Ages 8–9
© A & C BLACK

Rock, paper, scissors

Your teacher will explain the rules of 'Rock, paper, scissors'.

> Rock beats scissors, scissors beat paper and paper beats rock.

☆ Play the game with a partner.

☆ In the first column of the chart, write your name and the name of your partner.

☆ Keep a tally of who wins each game.

☆ Finally, add up the total number of wins for each player and complete the chart.

Name	Tally of wins	Total number of wins

• **Use the tally chart to answer these questions.**

1 How many games were played altogether? ＿＿＿

2 Who won the most games? ＿＿＿＿＿＿＿＿＿＿＿

3 What is the difference between the scores? ＿＿＿

NOW TRY THIS!

• **Write each of these tally results in numbers.**

Tally of wins	Total number of wins
卌 卌 卌 卌 卌 卌 |||	
卌 卌 卌 卌 ||||	
卌 卌 卌 卌 卌 卌 卌 卌 |||	

20

Teachers' note Ensure that the children are familiar with the rules of 'Rock, paper, scissors', which should be played in pairs (this could be done in mixed-ability pairs). Give the children a time limit or specify a number of games to play. A stopwatch could be used if the children are playing for a set number of minutes.

100% New Developing Mathematics
Handling Data: Ages 8–9
© A & C BLACK

Hen diagram

Farmer Feather keeps large and small hens which are speckled, black, or white. The hens have been sorted into the Venn diagram.

- Look carefully at the four labels.

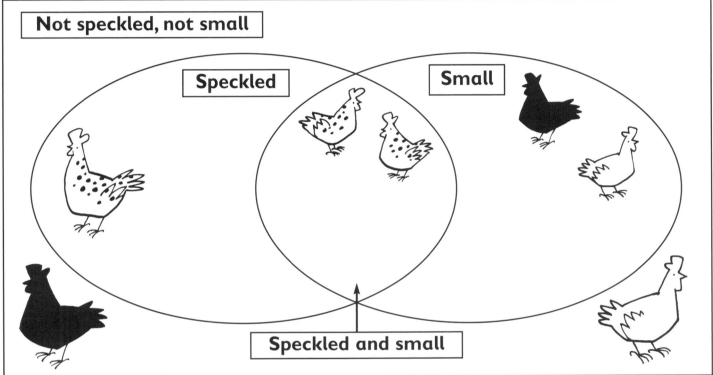

Not speckled, not small

Speckled

Small

Speckled and small

- **Complete this table using the data in the Venn diagram.**

Type of hen	Total number
Speckled	
Small	
Speckled and small	
Not speckled, not small	

NOW TRY THIS!

- **Draw these hens on the Venn diagram.**

Teachers' note Ensure that the children have a clear understanding that only two criteria are used on the Venn diagram, which are reflected in the labels. Explain that the diagram must be able to represent all data, including new data, for example a black and white hen.

100% New Developing Mathematics
Handling data: Ages 8–9
© A & C BLACK

1 Write each letter in the correct section of the Venn diagram.

One has been done for you.

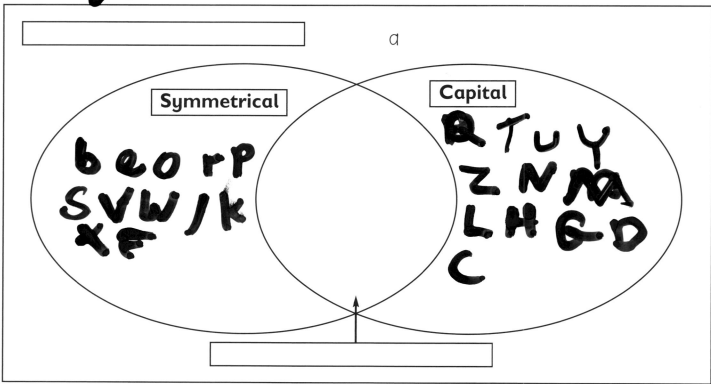

a

| Symmetrical | Capital |

b e o r p S V W J K X F

a

R T U Y Z N M L H G D C

2 There are 26 letters above. How many are:
  a capital? _____   **b** symmetrical? _____
3 How many letters are in the middle section? _____
4 How did you decide which letters to put in the middle section?

_____

5 There are six letters inside the rectangle and outside the circles. How would you describe this data? _____

NOW TRY THIS!

• **Fill in the empty labels to complete the Venn diagram.**

Teachers' note Ensure the children understand that the alphabet shows some capital and some lowercase letters. Some children might find it helpful to have a small mirror to help them decide whether the letters are symmetrical. Remind them that letters can have vertical or horizontal lines of symmetry.

100% New Developing Mathematics
Handling Data: Ages 8–9
© A & C BLACK

Monsters' tea party

This **Carroll diagram** shows data from the monsters' tea party.

	Chocolate cake	Not chocolate cake
Orange juice	Elmo, Nessie, Ziggy, Cleo, Dino, Percy, Mitzi	Cornelius, Minnie
Not orange juice	Taz, Wolfy, Slimy, Bogle	Howie, Stinker, Flower

1 How many monsters were at the tea party? ____

2 How many monsters had chocolate cake **and** orange juice? ____

3 How many monsters did **not** have orange juice? ____

4 Who had orange juice but did not have chocolate cake?

_____

5 How many monsters had chocolate cake? ____

6 Did Taz have orange juice? Yes ☐ No ☐

NOW TRY THIS!

Do you like chocolate cake and orange juice?
- **Write your name on the diagram to show what you would have.**

Teachers' note Discuss the diagram before the children answer the questions. Ensure that they understand the layout. The same data could be presented on a Venn diagram and comparisons made between the two diagrams.

100% New Developing Mathematics
Handling data: Ages 8–9
© A & C BLACK

2s and 3s

1 Write each of these numbers in the correct section
of the Carroll diagram.

16 21 33 12 101 2 10 35 30 18 1 6 8 14

	Multiple of 2	Not a multiple of 2
Multiple of 3		
Not a Multiple of 3	16	

One has been done for you.

2 a How many numbers in the diagram are a multiple of 2
and a multiple of 3? ___

b Write them here: _____

3 How many numbers in the diagram are a multiple of 2 but **not**
a multiple of 3? ___

4 In a different colour, write two more numbers that you think
should be included in each section of the Carroll diagram.

NOW TRY THIS!

- **How could you check whether you have placed the numbers correctly?**
- **Talk to a partner about your ideas.**

Teachers' note Children who find it difficult to remember their 2 times- and 3 times-tables could use tables grids to help them. As a further extension, the children could draw a Venn diagram of the same data.

100% New Developing Mathematics
Handling Data: Ages 8–9
© A & C BLACK

You need: two blank Venn diagrams, two blank Carroll diagrams, some scissors and glue.

☆ Cut out the sweets, one set at a time.

☆ Sort each set of sweets onto a Venn diagram or a Carroll diagram.

☆ Try to find four **different** ways to sort the sweets.

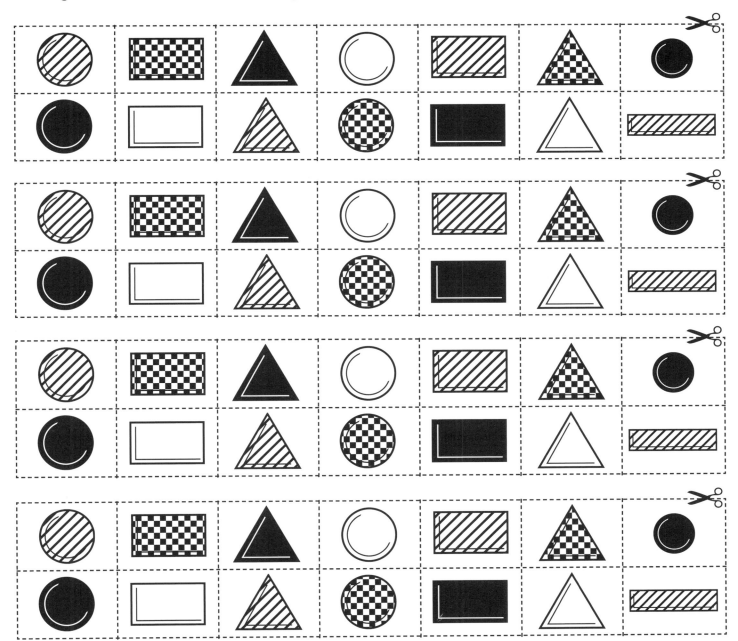

Teachers' note Use in conjunction with page 59, Blank Venn diagram, and page 60, Blank Carroll diagram. There are more than four ways to sort the sweets so you could demonstrate one way of sorting first, for example 'triangular sweets' and 'checked sweets'. Encourage the children to think about how they can check that they have found all the ways to sort the sweets.

100% New Developing Mathematics
Handling data: Ages 8–9
© A & C BLACK

Minibeasts

This pictogram shows how many minibeasts were found during a class investigation.

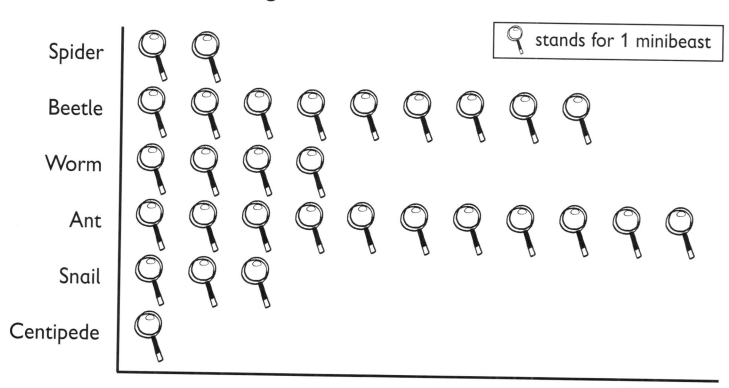

stands for 1 minibeast

1 How many spiders were found? ____

2 How many ants were found? ____

3 How many more beetles than worms were found? ____

4 How many fewer spiders than snails were found? ____

5 There were **exactly** four of which minibeast? _____

6 How many minibeasts were found altogether? ____

NOW TRY THIS!

Kajwast found two snails stuck to the lid of her container.
- **Add this data to the pictogram.**
- **On the back of this sheet, make up a question about the new pictogram.**

Teachers' note Talk about the pictogram before the children answer the questions. Explain that the symbol of the magnifying glass is used to represent any minibeast.

100% New Developing Mathematics
Handling Data: Ages 8–9
© A & C BLACK

Dani sorted out her sock drawer and
recorded the results in a pictogram.

| stands for 2 socks

Stripy

Spotty

White

Black

Pink

Fluffy

1 How many pairs of stripy socks does Dani have? ____

2 How many pairs of white socks are there? ____

3 There are five pairs of which colour sock? ____

4 a What does the data tell us about the number of fluffy socks?

_____

b What might be the reason for this? _____

5 a How many **pairs** of socks does Dani have? _____

b How many socks does she have altogether? ____

NOW TRY THIS!

Dani finds six spotty socks in another drawer.

• **Add these to the pictogram.**

Teachers' note Ensure that the children appreciate that Dani's socks are either stripy or spotty or white etc. Draw attention to the scale to ensure that the children are clear that one sock icon represents two socks or a pair of socks. The 1·5 fluffy socks in the pictogram could stimulate discussion in pairs or in small groups.

100% New Developing Mathematics
Handling data: Ages 8–9
© A & C BLACK

27

Summer fête

This pictogram shows the number of people who took part in different activities at the school fête.

	✗ stands for 10 people

Coconut shy	✗ ✗ ✗
Tug-of-war	✗ ✗ ✗ ✗ ✗
Beat the goalie	✗ ✗ ✗ ✗ ✗ ✗
Skittles	✗ ✗ ✗ ✗ ✗ ✗ ✗ ⌐
Hook-a-duck	✗ ✗ ✗ ✗ ⌐

1 How many people had a go on the Coconut shy?____
2 How many people took part in Hook-a-duck? ____
3 Fifty people took part in which activity?

_____

4 How many more people took part
   in Skittles than Hook-a-duck? ____
5 How many fewer people took part
   in Beat the goalie than Skittles? ____

NOW TRY THIS!

Thirty-five people guessed how many sweets there were in a jar.

• **How would you show this on the pictogram?**

Guess how many [   ]

Teachers' note Ensure that the children understand the vocabulary for the different activities. Draw attention to the symbol and explain that each one represents ten people.

100% New Developing Mathematics
Handling Data: Ages 8–9
© A & C BLACK

Super Song Star

**Super Song Star is a competition for school children.
The pictogram shows the number of votes
(to the nearest 50) for each act that got to the final.**

🎤 stands for 100 votes

Just So

Milly & Molly

Cool Kat

The Magnets

Boy Band

1 Which act won the Super Song Star? _____

2 Which act got 650 votes? _____

3 How many votes did Boy Band get? _____

4 How many votes did Cool Kat get? _____

5 How many more votes did Just So get than Cool Kat? _____

6 How many fewer votes did The Magnets get than
Milly & Molly? _____

NOW TRY THIS!

• **How many votes were there altogether
(to the nearest 50)?** _____

Teachers' note Ensure the children understand that the microphone symbol represents 100 votes and the half-microphone, 50 votes. A calculator may be needed for the extension activity. The children could use this data to draw a bar chart with a particular focus on choice of scale.

**100% New Developing Mathematics
Handling data: Ages 8–9
© A & C BLACK**

Stamp about: 1

- ## Work with a partner.

You need the envelopes cut from Stamp about: 2.

- ## Using an atlas or the Internet to help you, sort the envelopes into post for the UK, the rest of Europe, or the rest of the world.

- ## Construct a table and a pictogram to show this data.

Where is the letter going?	Total number of letters
UK	
Rest of Europe	
Rest of the world	

A pictogram to show the number of letters sent to the UK, the rest of Europe, and the rest of the world.

	stands for 2 letters

NOW TRY THIS!

- ## Make sure you have labelled your pictogram.
- ## How many letters were sent altogether? _____

Teachers' note Use in conjunction with page 31. Stamp about: 2. Access to atlases, maps, or the Internet will be needed to sort the data into destination categories. If desired, the number of letters that are represented by the symbol could be deleted and the children asked to say how many units they think it should represent.

100% New Developing Mathematics
Handling Data: Ages 8–9
© A & C BLACK

Stamp about: 2

● **Cut out the envelopes.**

Naples Italy	Cambridge Cambridgeshire	Prague Czech Republic	Kampala Uganda	Miami Florida USA	Winchester Hampshire
Paris France	Abingdon Oxfordshire	Wellington New Zealand	Moscow Russia	Geneva Switzerland	Cairo Egypt
Le Mans France	New York USA	Santa Lapland Finland	Timbuktu Mali	Salzburg Austria	Florence Italy
Mumbai India	Newport Isle of Wight	Rio de Janeiro Brazil	Copenhagen Denmark	Gunjur The Gambia	Madrid Spain
Sydney Australia	Leeds W. Yorkshire	Suva Fiji	Hereford Herefordshire	Cape Town South Africa	Belfast N. Ireland
Warsaw Poland	Perth Australia	Aberdeen Scotland	Tokyo Japan	Arillas Corfu	Liverpool Merseyside
Bangkok Thailand	Agra India	Frankfurt Germany	Cardiff Wales	Lima Peru	Chicago USA

Teachers' note Use in conjunction with page 30, Stamp about: 1.

100% New Developing Mathematics
Shapes and Measures:
Ages 5–6
© A & C BLACK

Mr Folly's lollies

This tally chart shows how many lollies Mr Folly sold during one week.

Monday	Tuesday	Wednesday	Thursday	Friday	Saturday	Sunday
|||| |||| |||| |||| ||||	|||| |||| ||||	||||	|||| ||||	|||| |||| |||| |||| |||| ||||	|||| |||| |||| |||| |||| |||| |||| |||| |||| |||| ||||	|||| |||| |||| |||| |||| |||| |||| ||||

• **Use the data to construct a pictogram.**

Think about how many lollies your picture will stand for.

A pictogram to show the number of lollies sold in one week

🍦 stands for ☐ lollies

Monday	
Tuesday	
Wednesday	
Thursday	
Friday	
Saturday	
Sunday	

NOW TRY THIS!

• **Why do you think that the number of lollies sold on Wednesday was so low?**

Talk to a partner about your ideas.

Teachers' note The children could be given some paired discussion time to consider how many lollies will be represented by the symbol used in the pictogram.

100% New Developing Mathematics
Handling Data: Ages 8–9
© A & C BLACK

After-school sports

A new after-school sports club is to be set up.
This bar chart shows how many children
voted for each sports club.

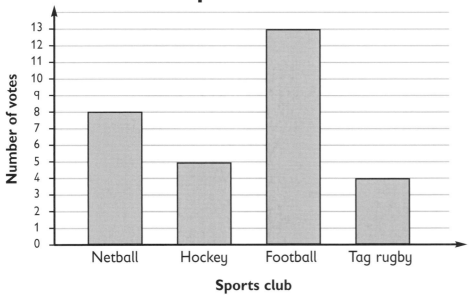

1 Which is the most popular choice of sport? _____

2 a From looking at the results, which sports club do you
think should be set up? _____

   b What are your reasons for this? (Talk to a partner about your ideas.)

   c How many votes did this sport get? ____

   d How many children did **not** vote for this club? ____

3 Have you changed your mind about which club should
be set up? Yes ☐   No ☐   (Talk to a partner about your reasons.)

NOW TRY THIS!

• **What would you suggest the school
does to solve this problem?**

(Talk to a partner about your ideas.)

Teachers' note The emphasis here is on analysing the data to solve a problem. The children need
to consider the proportion of votes that a particular sport did and did not get.

100% New Developing Mathematics
Handling data: Ages 8–9
© A & C BLACK

Splish, splash

This table shows the number of babies in the small pool on each day of the week.

Day	Number of babies
Monday	14
Tuesday	10
Wednesday	7
Thursday	5
Friday	16
Saturday	19
Sunday	9

- **Construct a bar chart to show this data.**

A bar chart to show the number of babies in the small pool in one week

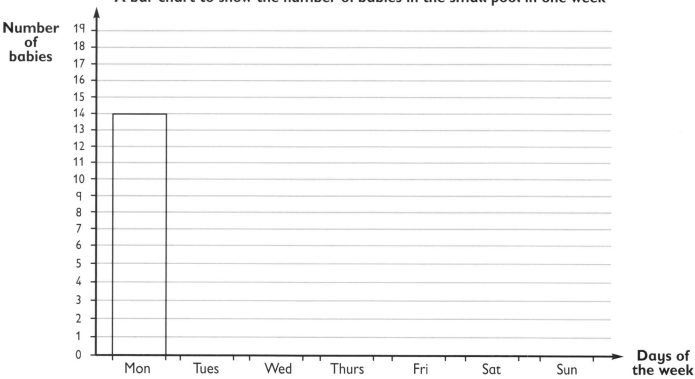

Number of babies (y-axis, labelled 0 to 19)

Days of the week (x-axis: Mon, Tues, Wed, Thurs, Fri, Sat, Sun)

NOW TRY THIS!

- **On a separate piece of paper, write four questions to ask a partner about the data in your chart.**

Teachers' note Ensure the children understand that the height of each bar shows how many babies were in the small pool that day.

100% New Developing Mathematics
Handling Data: Ages 8–9
© A & C BLACK

Family zoo

This bar chart shows the number of different pets owned by the children in Class 4.

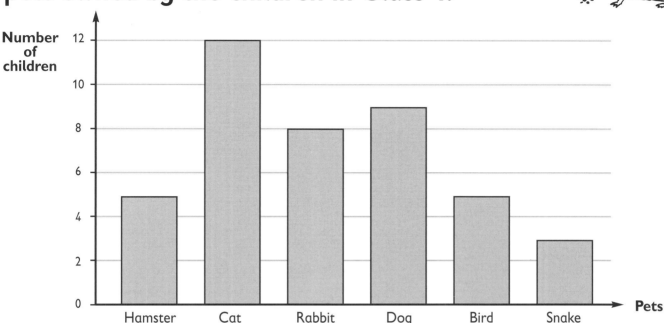

1 How many pets are owned altogether? _____

2 How many four-legged pets are there? _____

3 Which is the most popular type of pet? _____

4 How many fewer birds than cats are there? _____

5 Which two pets are equally popular? _____ and _____

6 How many more rabbits than snakes are there? _____

7 There are 30 children in Class 4. How can you explain your answer to question 1?

Talk to a partner about your ideas.

NOW TRY THIS!

- If you have any of these pets, add them to the bar chart.

Teachers' note Question 7 (and the page title as a clue) can support discussion about whether some children own more than one sort of pet, whilst others may have no pets at all. As this is the case, it is not possible to say how many people took part in the survey.

100% New Developing Mathematics
Handling data: Ages 8–9
© A & C BLACK

What a smoothie!

This table shows the number of fruits needed to make different smoothies.

Type of smoothie		Number of fruits needed
Pineapple		3
Peach		7
Banana		8
Strawberry		10
Mango		2

• Use this data to complete the bar chart.

Don't forget the labels on the axes.

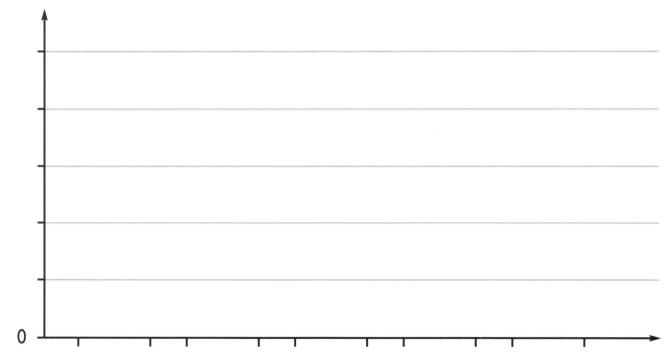

A bar chart to show the number of fruits needed to make different smoothies

Number of fruits needed

0

Type of smoothie

NOW TRY THIS!

• **Talk to a partner:**

a Did you both use the same scale? Yes ☐ No ☐

b What made you decide to use this scale?

Teachers' note If preferred, the children could draw the fruit on the horizontal axis, rather than writing the names of the fruit. The extension activity requires the children to work in pairs or small groups.

100% New Developing Mathematics
Handling Data: Ages 8–9
© A & C BLACK

Robot sort

This table shows what the robots on sale in Techno-World can do.

What robot can do	Number of robots
Spin	3
Walk	8
Talk	14
Grab	5

- **Complete this horizontal bar chart to show the information from the table.**

Label the axes first, then draw in the missing bars.

A bar chart to show what the robots on sale at Techno-World can do

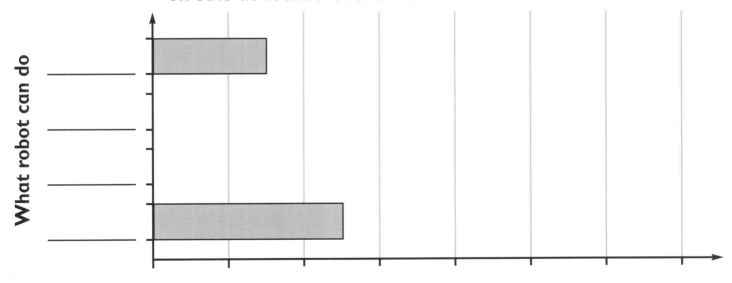

What robot can do

Number of robots

NOW TRY THIS!

- **How did you know which scale to use?**

Talk to a partner.

Teachers' note Check that the children appreciate that a robot can spin **or** walk **or** talk **or** grab; it cannot have more than one function. Ensure the children understand that the bar chart shows the same data as the table. Discuss what scale would be most appropriate for the horizontal axis.

100% New Developing Mathematics
Handling data: Ages 8–9
© A & C BLACK

Golden Giants

This table shows how many games each team won in a local basketball league.

• **Construct a vertical bar chart to show this data. Use a scale in steps of 2.**

Team		Number of games won
Flying Stars		13
Barracudas		8
Wild Cats		11
Golden Giants		17
Stingrays		10
Raptors		16

Use the worksheet called Blank vertical bar chart.

Remember to label the axes of your bar chart.

Remember to give your bar chart a title.

• **Use your bar chart to answer these questions.**

1 Which team won the league? _____

2 Which team won the fewest games? _____

3 How many games did the Stingrays win? _____

4 Which team won 11 games? _____

NOW TRY THIS!

A player from the Golden Giants has been asked to play for the Wild Cats.

• **Do you think he should transfer to the Wild Cats? Why do you think this?**

Talk to a partner about your ideas.

Teachers' note The children will need a copy of page 61: Blank vertical bar chart.

100% New Developing Mathematics
Handling Data: Ages 8–9
© A & C BLACK

Cool shades

Rupa took photos of some spectators at the outdoor tight-rope walking event. It was a very sunny day so they were all wearing shades.

- Fill in the totals: 🕶 ☐   🕶 ☐   🕶 ☐   🕶 ☐
- Construct a horizontal bar chart to show the number of people wearing each shape of shades.

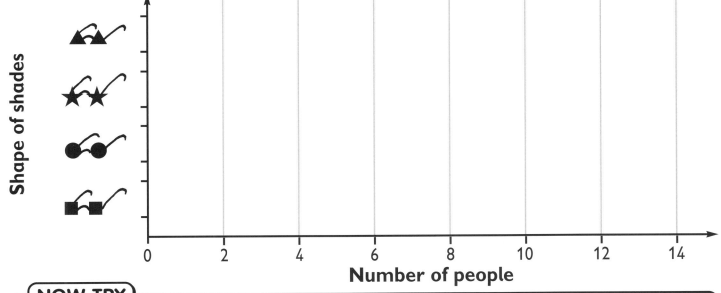

NOW TRY THIS!

a How many people wore triangular-shaped or star-shaped shades? ____

b What was the most popular shape of shades? _____

Teachers' note When constructing their bar charts, some children may find it easier to use the blank horizontal bar chart on page 62. ICT could also be used. Half the class could construct a vertical bar chart or pictogram for class evaluation of a range of representations at the end of the lesson.

100% New Developing Mathematics
Handling data: Ages 8–9
© A & C BLACK

Beachcomber

This table shows the things that Alice found on the beach.

Things Alice found		Number found
Crabs	🦀	9
Shells	🐚	27
Shiny pebbles		15
Starfish	⭐	4
Ice lolly sticks		5

Talk to a partner about what scale to use.

- **Present this data in a bar chart.**

My scale will go up in intervals of ☐

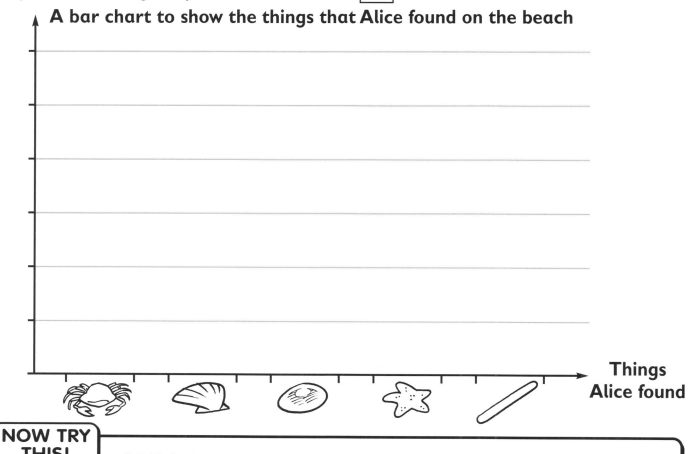

A bar chart to show the things that Alice found on the beach

Number found

Things Alice found

NOW TRY THIS!

- **Which bar did you find the most difficult to draw on the bar chart? Why?**

Talk to a partner.

Teachers' note The children should work in pairs or small groups to discuss an appropriate interval to use for their scale. Explain that all the data must be shown on the bar chart. Encourage the children to estimate as accurately as possible when plotting data with a total that falls between the intervals.

100% New Developing Mathematics
Handling Data: Ages 8–9
© A & C BLACK

Bounce 4 charity

William and his friends organised a sponsored trampoline event called 'Bounce 4 charity.' This bar chart shows the amount of money raised by each class.

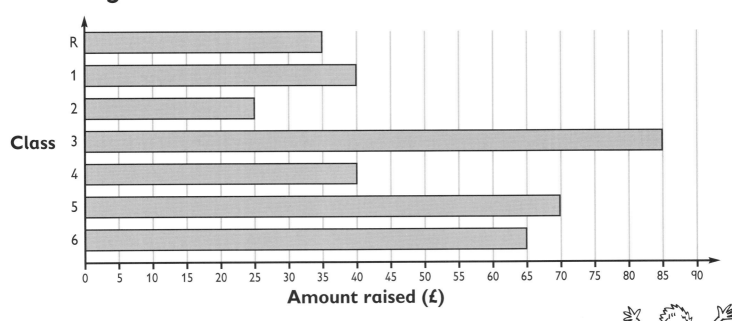

Class (vertical axis): R, 1, 2, 3, 4, 5, 6

Amount raised (£) (horizontal axis): 0, 5, 10, 15, 20, 25, 30, 35, 40, 45, 50, 55, 60, 65, 70, 75, 80, 85, 90

1 How much money did Class 5 raise? _____

2 Which class raised £25? _____

3 a Which two classes raised the same amount of money? _____ and _____

b How much did these two classes raise altogether? _____

4 a What do you notice about the amount raised by Class 3?

b What might explain this?

Talk to a partner about your ideas.

NOW TRY THIS!

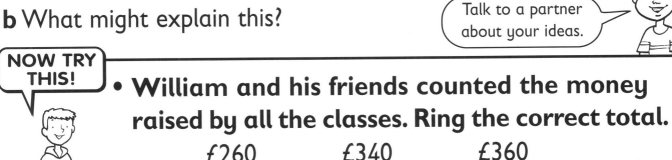

- **William and his friends counted the money raised by all the classes. Ring the correct total.**

£260   £340   £360

Teachers' note Discuss the scale used for the bar chart. The children should discuss their ideas for question 4 with a partner before a whole-class discussion.

100% New Developing Mathematics
Handling data: Ages 8–9
© A & C BLACK

Recycling

This bar chart shows how many kilograms of each material were collected for recycling from Reuze Road during one month.

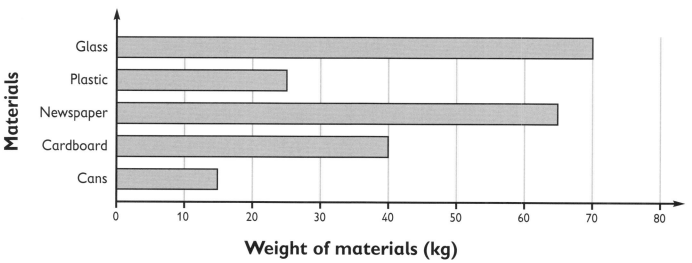

Weight of materials (kg)

1 How many kilograms of cardboard were recycled? _____ kg

2 Which material weighed 15 kg? _____

3 What was the total weight of newspaper and cardboard? _____ kg

4 Reuze Road's target weight for recycling material during one month is 200 kg.
Did they reach their target? Yes ☐ No ☐

NOW TRY THIS!

Ursula says: (Not many people recycle plastic.)

- **Do you think this is true ☐ or false ☐?**
- **Why do you think this?**

(Talk to a partner.)

Teachers' note Ensure the children are clear that the bar chart shows the mass of different materials, rather than individual pieces. Data showing local recycling habits could be used in the lesson and the children asked to present this information as a bar chart.

100% New Developing Mathematics
Handling Data: Ages 8–9
© A & C BLACK

Adventure holiday

There is a choice of activities on the last day of an adventure holiday. This tally chart shows how many children chose each activity.

Activity		Tally
Sailing	⛵	IIII IIII IIII
Horse riding	🐎	IIII III
Mountain biking	🚲	IIII
Canoeing	🛶	IIII IIII I
Grass sledging	🛷	IIII II

1 Construct a **horizontal** bar chart to show the results.

Use the worksheet called Blank horizontal bar chart.

• **Use your bar chart to answer these questions.**

2 **a** How many people chose horse riding? ____

b How many people chose a water-based activity? ____

c How many more people chose sailing than grass sledging? ____

NOW TRY THIS!

• **How many children chose to do an activity on the last day of the adventure holiday?** ____

Teachers' note The children will need a copy of page 62, Blank horizontal bar chart.

100% New Developing Mathematics
Handling data: Ages 8–9
© A & C BLACK

Odd one out: 1

In each set, two of the charts show the same data and one shows different data.

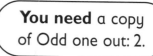
You need a copy of Odd one out: 2.

- **Which is the odd one out in each set? Put a tick in the box.**

Set 1 — Number of pets sold in one day at Pets 'R' Us

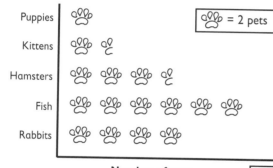

Pet: Puppies, Kittens, Hamsters, Fish, Rabbits

🐾 = 2 pets

Number of pets ☐

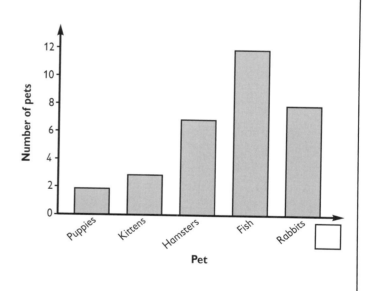

Pet	Tally
Puppies	II
Kittens	III
Hamsters	IIII I
Fish	IIII IIII II
Rabbits	IIII II

☐

NOW TRY THIS!

- **Use a coloured pencil to make changes to the odd one out in each set so that all three charts show the same data.**

Teachers' note Use in conjunction with page 45, Odd one out: 2. The children will need to be familiar with interpreting and constructing tally charts, pictograms and bar charts. Encourage them to use a ruler to read across from the bar to the numbered axis. Coloured pencils are needed for the extension activity.

100% New Developing Mathematics
Handling Data: Ages 8–9
© A & C BLACK

Odd one out: 2

You need a copy of Odd one out: 1.

Set 2 — Number of ice-creams sold in one day at Nice 'n Icy

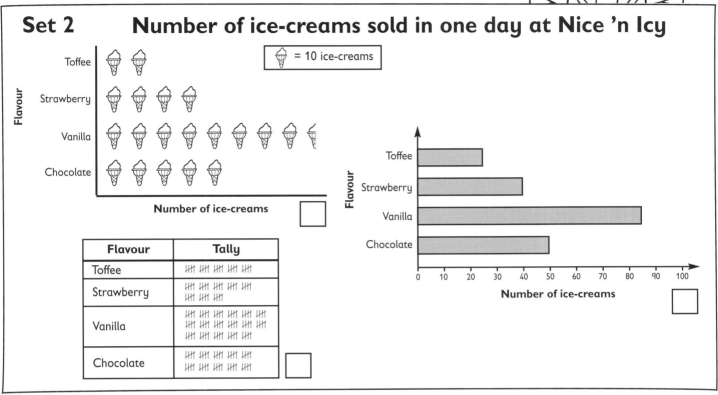

Flavour	Tally
Toffee	卌 卌 卌 卌 卌
Strawberry	卌 卌 卌 卌 卌 卌 卌 卌
Vanilla	卌 卌 卌 卌 卌 卌 卌 卌 卌 卌 卌 卌 卌 卌 卌 卌
Chocolate	卌 卌 卌 卌 卌 卌 卌 卌 卌 卌

Set 3 — Number of minibeasts seen in one day

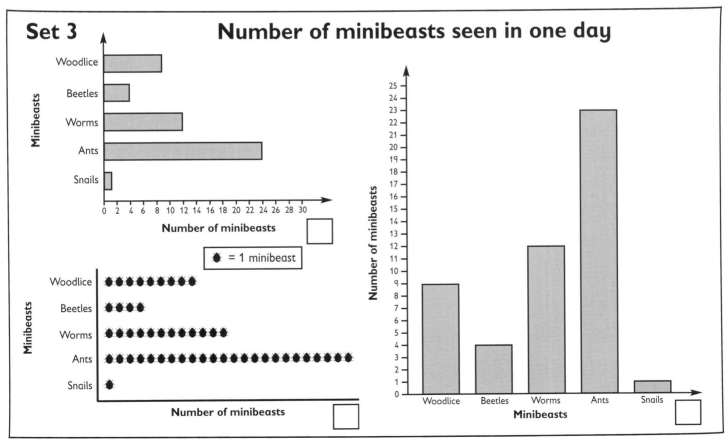

Teachers' note Use in conjunction with page 44, Odd one out: 1.

100% New Developing Mathematics
Handling data: Ages 8–9
© A & C BLACK

You need a copy of Game, set and match: 2.

- **Look at the four descriptions below.**
- **Match each description to the chart that you think represents the data.**

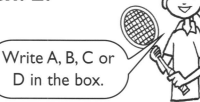
Write A, B, C or D in the box.

1 James and his five friends from football had a penalty shoot-out. They had eight shots each.

2 Six local schools took part in a netball championship. Each team played five games.

3 A tennis competition was held where each person played ten games. Four players took part.

4 Four wrestlers entered the annual wrestling match. Each wrestler took part in eight matches.

- **Compare your answers with those of a partner.**
- **Explain to your partner why you think your answers are correct.**

NOW TRY THIS!

- **Fill in the missing labels on each chart.**

Don't forget to give each chart a title.

Teachers' note Use in conjunction with page 47, Game, set and match: 2. There is one correct representation for each scenario. Ensure that the children appreciate that the charts on page 47 show the games and matches won and the goals scored. Encourage the children to explain why they think the other three representations cannot be correct, as well as giving a rationale for their chosen one.

100% New Developing Mathematics
Handling Data: Ages 8–9
© A & C BLACK

Game, set and match: 2

You need a copy of **Game, set and match: 1.**

• Look carefully at the charts.

A

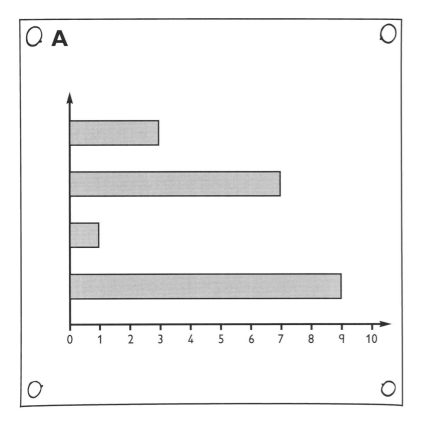

B

	Tally
	||||
	ⅢⅠ |
	ⅢⅠ
	|

C

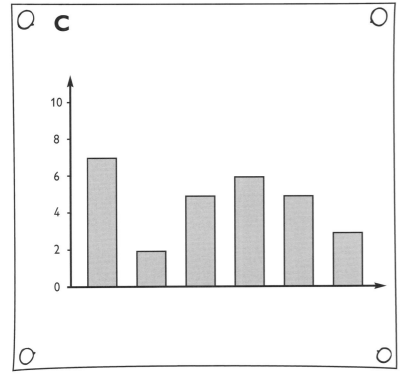

D

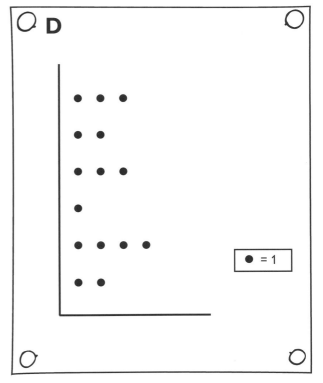

● = 1

Teachers' note Use in conjunction with page 46, Game, set and match: 1.

100% New Developing Mathematics
Handling data: Ages 8–9
© A & C BLACK

47

Missing data

A computer virus has deleted some of the data from this table and bar chart.
- Fill in the missing data.

You will need to piece together the information from the table and the bar chart.

A table and bar chart to show the number of individual pieces of artefact dug up at a recent excavation

Type of artefact	Number of pieces found
Animal bones	
	95
Glass	
Woven fabric	40
	20
Total number of pieces found	290

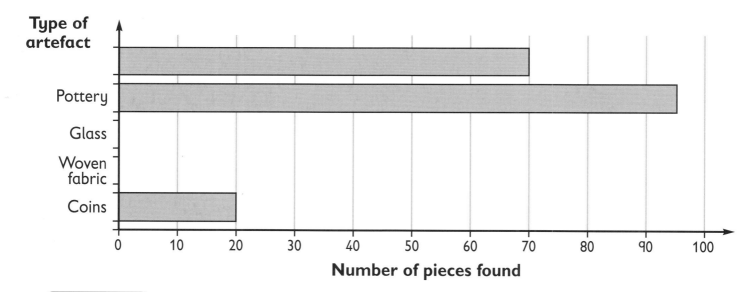

Number of pieces found

NOW TRY THIS!

- **Which was the most difficult data to find?**
- **What did you do to find it?**

Talk to a partner.

Teachers' note Ensure the children understand that all the data can be completed by piecing together the information from the table and the bar chart. Encourage the children to persevere and, if necessary, share their ideas with a partner if they cannot readily find all the answers.

**100% New Developing Mathematics
Handling Data: Ages 8–9
© A & C BLACK**

Top of the mountain: 1

• Play this game with a partner.

☆ Shuffle the cards and put them in a pile face down.

☆ Take turns to roll the dice and move your counter.

☆ If you land on a slimy square, pick up a card.

☆ If you get the answer right, slide forward one space.

☆ If you get it wrong, stick where you are.

☆ Return your card to the bottom of the pile.

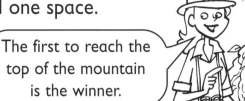

You need the cards cut from Top of the mountain: 2, a dice and two counters.

The first to reach the top of the mountain is the winner.

Teachers' note Use in conjunction with page 50, Top of the mountain: 2. The game board could be used on the interactive whiteboard for the class to play in teams, or it could be enlarged to A3 size for groups to play. Before they begin to play, check that children recognise that squares such as 3, 7 and 15 are 'slimy squares'.

100% New Developing Mathematics
Handling data: Ages 8–9
© A & C BLACK

Top of the mountain: 2

• Cut out the cards.

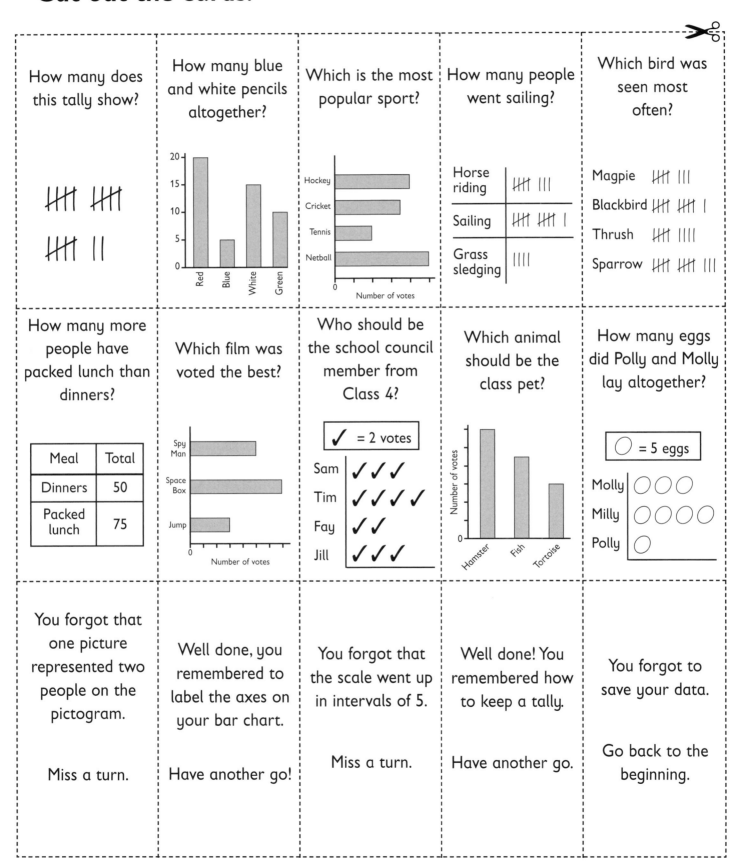

How many does this tally show?	How many blue and white pencils altogether?	Which is the most popular sport?	How many people went sailing?	Which bird was seen most often?
How many more people have packed lunch than dinners?	Which film was voted the best?	Who should be the school council member from Class 4?	Which animal should be the class pet?	How many eggs did Polly and Molly lay altogether?
You forgot that one picture represented two people on the pictogram. Miss a turn.	Well done, you remembered to label the axes on your bar chart. Have another go!	You forgot that the scale went up in intervals of 5. Miss a turn.	Well done! You remembered how to keep a tally. Have another go.	You forgot to save your data. Go back to the beginning.

Teachers' note Use in conjunction with page 49, Top of the Mountain: 2. Additional cards could be written by the children and added to the game resource.

**100% New Developing Mathematics
Handling Data: Ages 8–9
© A & C BLACK**

Mix and match: 1

You need the cards cut from Mix and match: 2.

- **Sort the bar charts into pairs which show the same data.**
- **Record the letters of the bar charts that you think are pairs.**

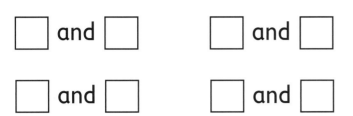

☐ and ☐   ☐ and ☐

☐ and ☐   ☐ and ☐

- **Compare your answers with those of a partner.**
- **Did you get the same answers?** Yes ☐ No ☐
  If not, check your answers.

- **How did you decide which bar charts showed the same data?** _____

_____

_____

- **Choose one pair of bar charts. Explain how you know this is a pair?** _____

_____

_____

NOW TRY THIS!

- **Construct a tally chart to show the data from one pair of cards.**
- **Ask a partner to work out which pair it matches.**

Teachers' note Use in conjunction with page 52, Mix and match: 2. Check that children understand that the pairs of bar charts show the **same data** but represent it differently.

100% New Developing Mathematics
Handling data: Ages 8–9
© A & C BLACK

Mix and match: 2

- **Cut out the cards below.**
- **Match the pairs of data and record on Mix and match: 1.**

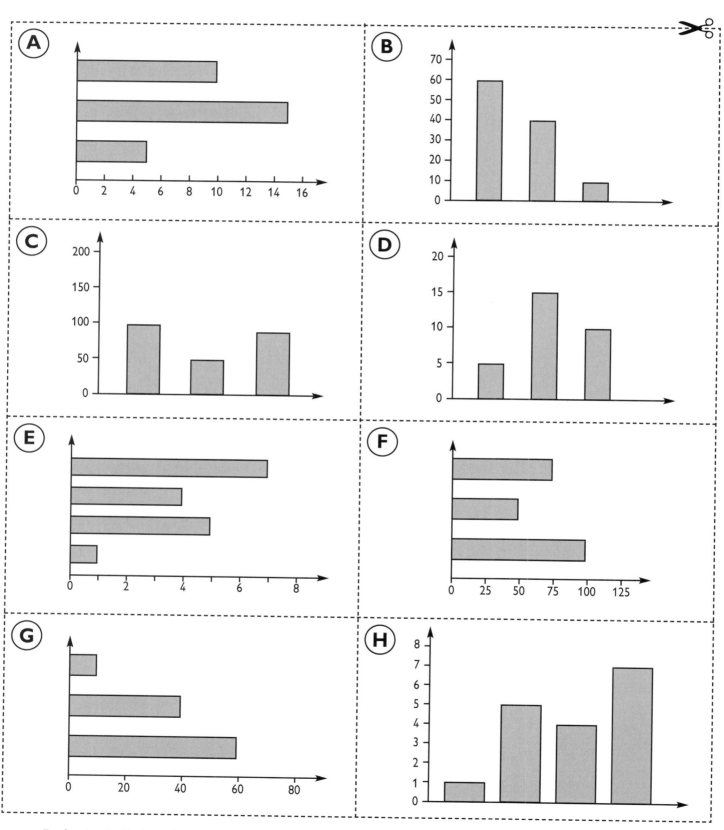

Teachers' note Use in conjunction with page 51. Mix and match: 1.

52

100% New Developing Mathematics
Handling Data: Ages 8–9
© A & C BLACK

Scaly fish: 1

This table shows the number of each type of fish in Mika's aquarium.

Type of fish		Number of fish
Rainbow fish		4
Zebra fish		9
Neon tetra		23
Angelfish		15

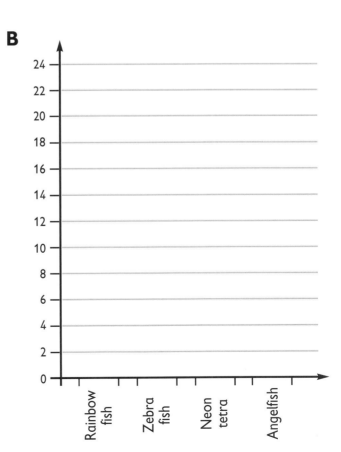

You need a copy of Scaly fish: 2.

Look carefully at the scales!

• **Construct four vertical bar charts to show this data.**

A

24, 23, 22, 21, 20, 19, 18, 17, 16, 15, 14, 13, 12, 11, 10, 9, 8, 7, 6, 5, 4, 3, 2, 1, 0

Rainbow fish   Zebra fish   Neon tetra   Angelfish

B

24, 22, 20, 18, 16, 14, 12, 10, 8, 6, 4, 2, 0

Rainbow fish   Zebra fish   Neon tetra   Angelfish

Teachers' note Use in conjunction with page 54, Scaly fish: 2. Draw the children's attention to the different scales on each of the four vertical axes. Ensure the children understand that they are plotting the **same** data on each bar chart.

100% New Developing Mathematics
Handling data: Ages 8–9
© A & C BLACK

C

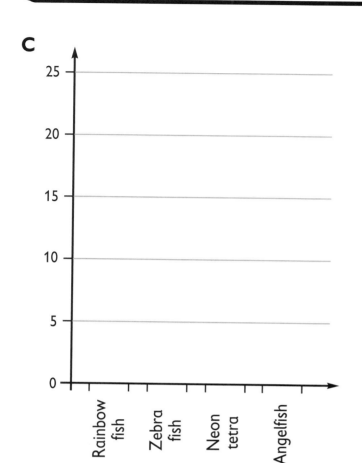

D

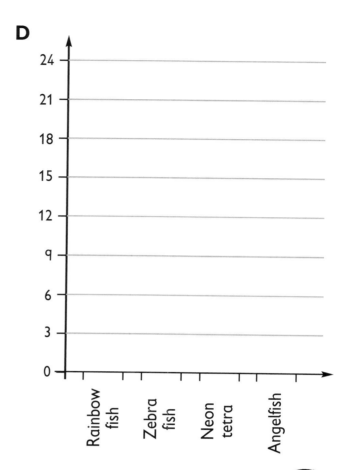

1 How did you work out where to draw the end of each bar?

Talk to a partner about how you did this.

2 Which bar chart was the easiest to draw? ___

3 Do you think the data looks the same in each bar chart?

Talk to a partner about your ideas.

NOW TRY THIS!

- **Which bar chart do you think shows the data most clearly?** ___
- **Why do you think this?**

Talk to a partner about your ideas.

Teachers' note Use in conjunction with page 53. Scaly fish: 1.

100% New Developing Mathematics
Handling Data: Ages 8–9
© A & C BLACK

Data investigation

Handling data cycle

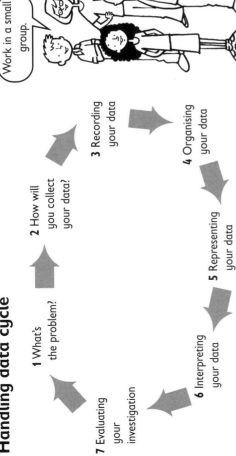

Work in a small group.

1 What's the problem?

2 How will you collect your data?

3 Recording your data

4 Organising your data

5 Representing your data

6 Interpreting your data

7 Evaluating your investigation

Question that we want to investigate

Members of the group

Teachers' note The children should work in a small group for the whole cycle of the handling data investigation.

100% New Developing Mathematics
Handling data: Ages 8–9
© A & C BLACK

55a

What's the problem?

• **Think of some questions that you could investigate and write them in the box.**

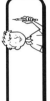

• **Talk to the others in the group about your ideas. Together, choose the question that the group would like to investigate.**

• **What do you think you will find out at the end of the investigation? Write your group's prediction below.**

• **What makes you think this?**

Teachers' note The initial generation of ideas should be done individually to ensure that everyone has had the opportunity to put their idea forward. Ensure that the prediction is used to guide the group's thinking, rather than to prescribe the outcome at this stage.

100% New Developing Mathematics
Handling data: Ages 8–9   55b
© A & C BLACK

How will you collect your data?

- **As a group, discuss which of the methods above you will use to collect the data. Ring the chosen method.**

Counting	Tally	Survey	Other idea
Timing	Questionnaire		

- **Explain why you think this is the best method.**

This is the best method to collect the data for this project because

- **Talk to your group and answer these questions.**

1 Who or where will you get your data from?

2 How will you make sure that you do not miss some data or collect the same data twice?

3 What have you done to make sure that you have a fair sample?

56a

Teachers' note This stage of the investigation follows work on different methods of data collection. Ensure that the children understand the strengths of each method and, as a class, discuss the nature of projects that would suit each one.

100% New Developing Mathematics
Handling data: Ages 8–9
© A & C BLACK

Record your data

- **Look at your investigation question.**
- **How will you record the data you need?**

- **Use this space to create a record sheet for the data.**

> You might write a questionnaire or draw a table or diagram.

56b

Teachers' note The children now plan their questionnaire or construct their table, entering any details they need to facilitate collection. The children could use ICT to create a clearer recording sheet which can be edited before use. Questionnaires may need photocopying.

100% New Developing Mathematics
Handling data: Ages 8–9
© A & C BLACK

Represent your data

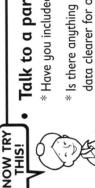

- **How are you going to represent your data? Ring your chosen idea.**

  Pictogram   Bar chart   Diagram

- **Give a reason why you have chosen to represent your data in this way.**

  _____

  > Think about what scale is best to use and remember to include a title and labels.

- **Represent your data in a clear way to show your results.**

NOW TRY THIS!

- **Talk to a partner about the following:**
  * Have you included titles and labels?
  * Is there anything you could do to make your data clearer for others to understand?

Teachers' note Individual children may choose to represent their data in different ways. Encourage this so that there is an opportunity for the group to compare different representations of the same data. If using ICT, ensure that the children are familiar with the appropriate programmes.

100% New Developing
Mathematics
Handling data: Ages 8–9
© A & C BLACK

Organise your data

You should now have collected all of your data.

- **Use this table to organise your results.**

> Are there any problems with the data you have collected?

> Do you need to check anything at this stage?

NOW TRY THIS!

- **Talk to the others in your group.**
- **Do you need to find out anything else before you can represent your data?**

Teachers' note Depending on the nature of the data collected, the children may need to organise their data before they decide how they want to represent it. This is of particular value if a survey or questionnaire has been used.

100% New Developing
Mathematics
Handling data: Ages 8–9
© A & C BLACK

Interpret your data

Talk to your group about your ideas.

• What does your data tell you?

• Write some questions about your data for other groups to answer.

• Look at the worksheet called **What's the problem?** Was your prediction correct? Yes ☐ No ☐

• Talk to someone from another group. Swap worksheets and try to answer the questions they have written about their results.

58a

Teachers' note The first question requires the children to talk about what their results show. Prompt questions, tailored to each group, could be given. Although linked back to their prediction, encourage them to consider the extent to which they were right or wrong.

100% New Developing
Mathematics
Handling data: Ages 8–9
© A & C BLACK

Evaluate your investigation

• Did you collect enough of the right sort of data to answer your investigation question? If not, what could you have done differently?

• Do you think you could have done anything better? Think about how you collected and represented your data.

• Have you got any more questions that you would like to investigate? This could be linked to your investigation, a friend's investigation or a new idea.

• What did other groups do well?
• Note down one or two ideas here:

58b

Teachers' note Ensure there is enough time for the children to evaluate the whole handling data cycle so that they can see that it ends with consideration of the extent to which their question has been answered.

100% New Developing
Mathematics
Handling data: Ages 8–9
© A & C BLACK

Blank Venn diagram

Title _____

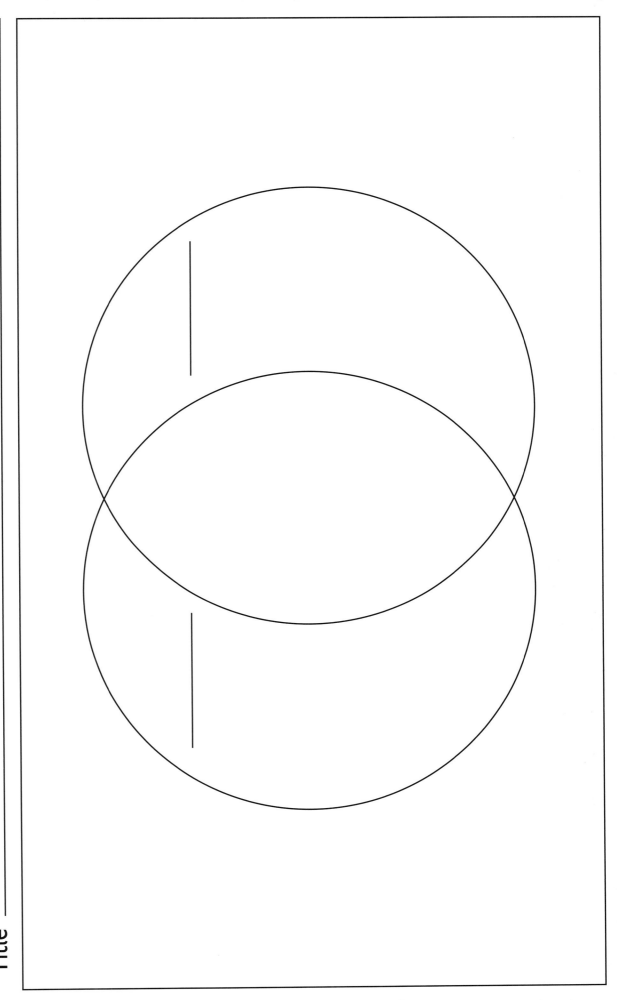

Teachers' note You could add a title and labels for the rings, or you could ask the children to complete these themselves.

100% New Developing Mathematics
Handling data:
Ages 8–9
© A & C BLACK

Blank Carroll diagram

Title

Teachers' note You could add a title and labels for the rows and columns, or you could ask the children to complete these themselves.

100% New Developing Mathematics
Handling data:
Ages 8–9
© A & C BLACK

Blank vertical bar chart

Title _____

_____

_____

Teachers' note You could add a title and labels for the axes, or you could ask the children to
complete these themselves.

100% New Developing Mathematics
Handling data: Ages 8–9
© A & C BLACK

61

Blank horizontal bar chart

Title

Teachers' note You could add a title and labels for the axes, or you could ask the children to complete these themselves.

62

100% New Developing Mathematics
Handling data:
Number: Ages 8–9
© A & C BLACK

Answers

p 13

a measuring: times taken by everyone in the class to run the same distance

b tally: number of people walking their dogs past the school in each hour of the day

c measuring: temperatures of different parts of the room

d counting: number of children aged 8 in the class and those aged 9

e counting or survey: name of favourite computer game for each child in the class

p 14

The femur is the longest bone in the body.

p 15

Type of jelly	Number of jellies
Rabbit jelly	4
Wobbly jelly	13
Brain jelly	8
Pumpkin jelly	3
Bear jelly	7
Total number of jellies	35

Now try this!

lime 13   strawberry 15   blackcurrant 7

p 17

1 8 cm   **2** March and November   **3** 4

4 a October **b** 4 cm   **5** Answers will vary.

Now try this!

87 cm

p 18

1 flip-flops 17   plimsolls 7   trainers 23

boots 1   wellies 4   school shoes 2

2 4   **3** 17   **4** plimsolls

5 a boots

b Because it was hot. Rationale: it was a summer fête. Lots of people were wearing flip-flops.

Now try this!

56 children went on the bouncy castle. This includes the two children who turned up wearing no shoes.

p 19

1 Chestnut 7   Dapple 12   Indigo Blue 4

Jasper 15   White Knight 0   Oyster Pearl 11

2 a Jasper   **b** 11

3 Oyster Pearl

4 Suggestions might include: he wasn't in the race; he couldn't run; he wasn't very good at jumping; he stayed behind to eat grass.

Now try this!

卌 II

p 20 Now try this!

33, 24, 43

p 21

Type of hen	Total number
Speckled	3
Small	4
Speckled and small	2
Not speckled, not small	2

Now try this!

small white: right circle

small speckled: intersection

large black: outside of circles

large speckled: left circle

large black and white: outside of circles

p 22

1

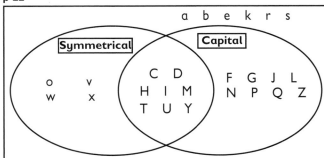

2 a 16

b 12

3 8

4 Responses might include knowledge of letters and use of the mirror/lines of symmetry.

5 Not symmetrical, not capital

Now try this!

Rectangle: Not symmetrical, not capital

Intersection: Symmetrical and capital

p 23

1 16   **2** 7   **3** 7

4 Cornelius and Minnie   **5** 11   **6** No

p 24

1

	Multiple of 2	Not a multiple of 2
Multiple of 3	12 30 18 6	21 33
Not a Multiple of 3	16 2 10 8 14	101 35 1

2 a 4   **b** 12, 30, 18, 6   **3** 5   **4** Answers will vary.

p 25

Ways of sorting include: triangular sweets/stripy sweets; circular sweets/black sweets; rectangular sweets/white sweets; circular sweets/triangular sweets, etc.

p 26

1 2   **2** 11   **3** 5

4 1   **5** worm   **6** 30

p 27

1 6

2 3

3 black

4 a There is an incomplete pair – there are 3 socks.

b Dani might have lost one.

5 a 21·5 pairs

b 43 socks

p 28

1 30   **2** 45   **3** Tug-of-war

4 30   **5** 15

Now try this!

p 29
1 Milly & Molly **2** The Magnets **3** 400
4 300 **5** 250 **6** 50

Now try this!
2600

p 30

Where is the letter going?	Total number of letters
UK	10
Europe	14
Rest of the World	18

Now try this!
42

p 32
It is likely that the symbol will represent 5 or 10 lollies sold.

p 33
1 football
2 a It is likely that the children will say football.
  b It is the most popular sport. Some might give the reason that it is their favourite sport.
  c 13
  d 17
3 There is not a correct answer here. The answer will depend on what children think. The important part is that they are able to give a reason for their decision.

Now try this!
Again, there is not a correct answer here.

p 35
1 42
2 34
3 cat
4 7
5 hamster and bird
6 5
7 A child may own more than one type of pet or no pets at all. Some children might own pets that are not listed above, for example guinea pigs. It is not possible to know how many people took part in the survey by looking at the data alone.

p 38
1 the Golden Giants **2** the Barracudas
3 10 **4** the Wild Cats

Now try this!
The data from the league results shows that the Golden Giants were better than the Wild Cats, so transferring teams would not be recommended. Children could put forward their own rationale.

p 39
triangular shades: 6 star-shaped shades: 2
round shades: 13 square shades: 9

Now try this!
a 8 **b** round

p 41
1 £70
2 Class 2
3a Class 1 and Class 4
  b £80

4 It was quite a lot more than any other class. It could be explained by there being more children in that class, that they out-bounced the other classes, or that their donors were more generous. Encourage the children to think of likely explanations.

Now try this!
£360

p 42
1 40 kg **2** cans **3** 105 kg **4** Yes

Now try this!
The data does not show how many of each item are recycled or how many people recycle materials. As plastic is lighter than some of the other resources, it appears as though not many people recycle it.

p 43
2 a 8 **b** 25 **c** 7

Now try this!
45

p 44–45
Set 1: the number of hamsters and rabbits needs correcting on the tally chart.
Set 2: toffee needs correcting on the pictogram.
Set 3: the number of ants needs correcting on the vertical bar chart.

p 46
1 C **2** D **3** A **4** B

p 48

Type of artefact	Number of pieces found
Animal bones	70
Pottery	95
Glass	65
Woven fabric	40
Coins	20
Total number of pieces found	290

Now try this!
The most difficult data to find is the number of pieces of glass. This data is found by calculating the difference between the total number of pieces found (290) and the totals that are known for each type of artefact.

p 50
Tally shows 17
20 blue and white pencils altogether
Netball is the most popular sport
11 people went sailing
Sparrows were seen most often
25 more people have packed lunch than school dinners
Space Box was voted best film
Tim should be the school council member
A hamster should be the class pet
Polly and Molly laid 4 eggs altogether

p 51
A and D B and G
C and F E and H